GARAM

가람 고급

가람 고급 (원제:GARAM NIVEAU EXPERT)

1판 1쇄 2019년 5월 20일
3쇄 2023년 1월 3일

지 은 이 람세스 분쾨사포
발 행 인 주정관
발 행 처 북스토리(주)
주 소 서울특별시 마포구 양화로 7길 6-16 서교제일빌딩 201호
대표전화 02-332-5281
팩시밀리 02-332-5283
출판등록 1999년 8월 18일 (제22-1610호)
홈페이지 www.ebookstory.co.kr
이 메 일 bookstory@naver.com

ISBN 979-11-5564-165-1 14410
979-11-5564-163-7 (세트)

※잘못된 책은 바꾸어드립니다.

프랑스를 강타한 새로운 두뇌 워밍업 수학 퍼즐

GARAM

가람 고급

람세스 분쾨사포 지음

북스토리

GARAM(가람)이란?

숫자 2개로만 이루어진, 간단한 연산에 기초한 논리적인 수학 퍼즐입니다. 연산의 결과는 한 자릿수 또는 두 자릿수가 될 수 있습니다.

■ □

■ 규칙

이미 채워진 수를 기초로 행과 열이 온전하게 연산이 성립하도록 빈칸에 숫자를 하나씩 넣습니다. 세로 연산에서 등호(=) 뒤에 빈칸 2개가 올 경우, 연산의 결과는 두 자릿수입니다(단, 두 자릿수는 0으로 시작하지 않습니다).

■ 팁

모든 문제의 정답은 하나뿐이므로 정답이라는 확신이 들 때 빈칸을 채워야 합니다. 다양한 경우의 수를 적어둔 다음 오답을 하나씩 지우며 정답을 가려내는 것도 좋은 방법이 될 수 있습니다. 별 한 개에서 다섯 개까지 점차적으로 난이도를 높여가며 150문제를 수록했기 때문에 차례대로 건너뛰지 말고 푸는 것이 좋습니다.

■ 예제

회색 빈칸을 먼저 채운 후, 다음의 힌트를 참고해서 퍼즐을 완성하세요.

① 7을 몇으로 곱하면 10~19 사이의 수가 나올까?

② 3단에서 2로 끝나는 수는?

③ 7을 몇으로 곱하면 20~30 사이의 짝수가 나올까?

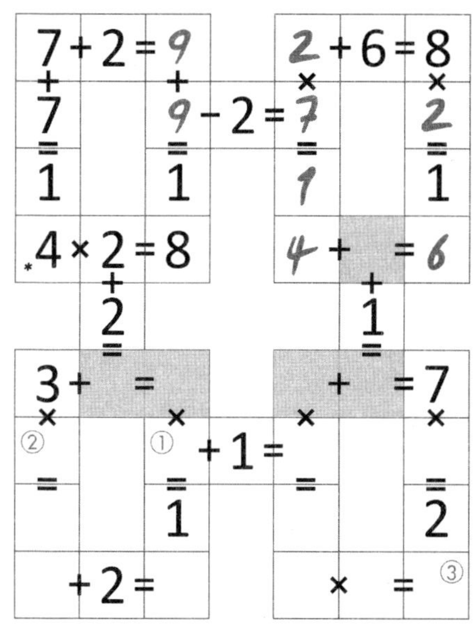

*모든 연산은 서로 연결되어 있으니 유의하세요!
가로로 4×2=8이라고 읽고, 세로로는 7+7=14라고 읽습니다.

이제 이해하셨나요?

■ □ ■ 수학 퍼즐 가람의 세계에 빠져보세요! ■ □ ■

001

★☆☆☆☆

+ 1 =

+ × − 5 = + − 2 = ×

= = = =

1 3

× = + =

− +

4

= =

− = + =

× × × ×

+ = 7

= = = =

3

× 5 = + 0 =

GARAM

002

★☆☆☆☆

− 6 =   + =

×   ×   ×   +

3   − =   7

=   =   =   =

3

× =   × =

+   −

5

=   =

+ =   + =

×   +   +   ×

4   − 2 =   3

=   =   =   =

− 1 =   + 6 =

003

★☆☆☆☆

+ 2 =　　　　+ 6 =

\+ 　+ 　× 　+

7 　− 　= 　5

= 　= 　= 　=

× 　= 　　× 　=

− 　　+

5

= 　　=

\+ 　= 　　− 　=

\+ 　+ 　× 　+

3 　+ 3 = 　5

= 　= 　= 　=

× 3 = 　　× 4 =

004

★☆☆☆☆

− 6 =   + 2 =

\+   ×   ×   +

3   + 2 =

=   =   =   =

1

× =   + =

−   +

2

=   =

\+ =   − =

\+   ×   ×   +

6   + =   4

=   =   =   =

\+ 8 =   − 4 =

005

★☆☆☆☆

+ 2 =   + 6 =

+   ×   +   +

−   =   5

=   =   =   =

4

+   =   ×   =

+   ×

3   1

=   =

−   =   +   =

+   ×   ×   +

3   +   =

=   =   =   =

2

+ 1 =   + 5 =

006

★☆☆☆☆

+ 5 = + 4 =

× + + ×

5 − =

= = = =

4

× = × =

− −

4

= =

+ = + =

× × × +

5 × 1 = 8

= = = =

4 4

× = × =

007

★☆☆☆☆

− = − 3 =

× × + ×

− 1 = 2

= = = =

3 3

− = + =

− +

2

= =

− = − =

+ × × ×

4 + = 8

= = = =

× 9 = − 7 =

008

★☆☆☆☆

− 1 = + =

\+ × × +

8 − = 3

= = = =

3

− = − =

− +

3

= =

\+ = − =

× + + ×

7 × 2 = 4

= = = =

− 0 = + 4 =

009

★☆☆☆☆

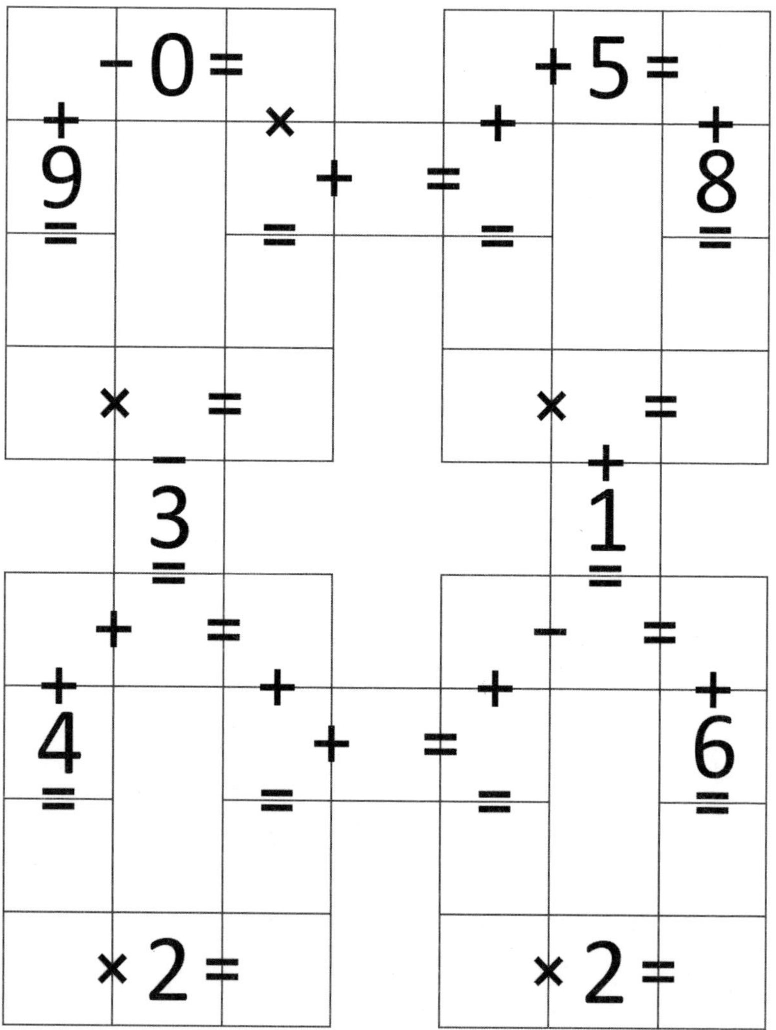

010

★☆☆☆☆

− 6 =   − =

×   ×   ×   ×

3   + =

=   =   =   =

2   3

+ =   − =

−   −

1   5

=   =

− =   − =

+   ×   ×   +

3   + =

=   =   =   =

4   2

− =   − 1 =

011

★☆☆☆☆

+ 2 =　　+ 2 =

× + − 3 = + + 9

= = = =

4

+ =　　× =

+　　+

=　　2

=

+ =　　+ =

× × − = × + 4

= = = =

3　　2

× 2 =　　× =

014

★☆☆☆☆

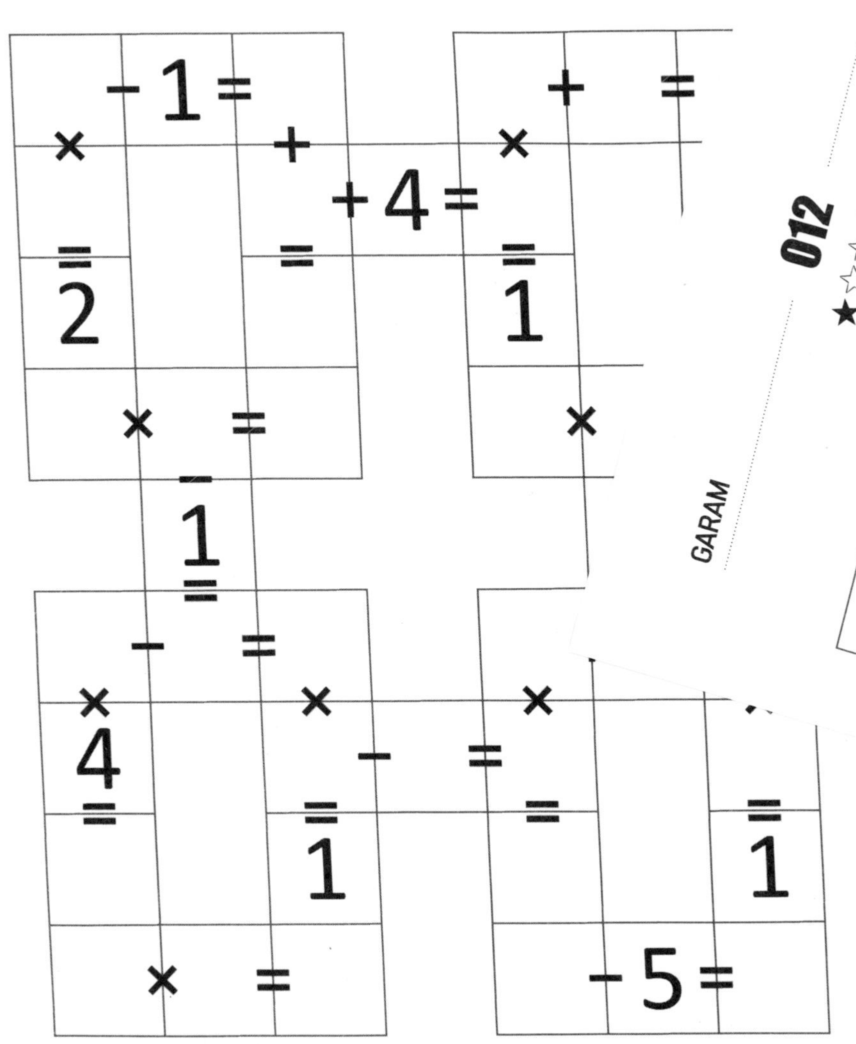

012

★☆☆☆☆

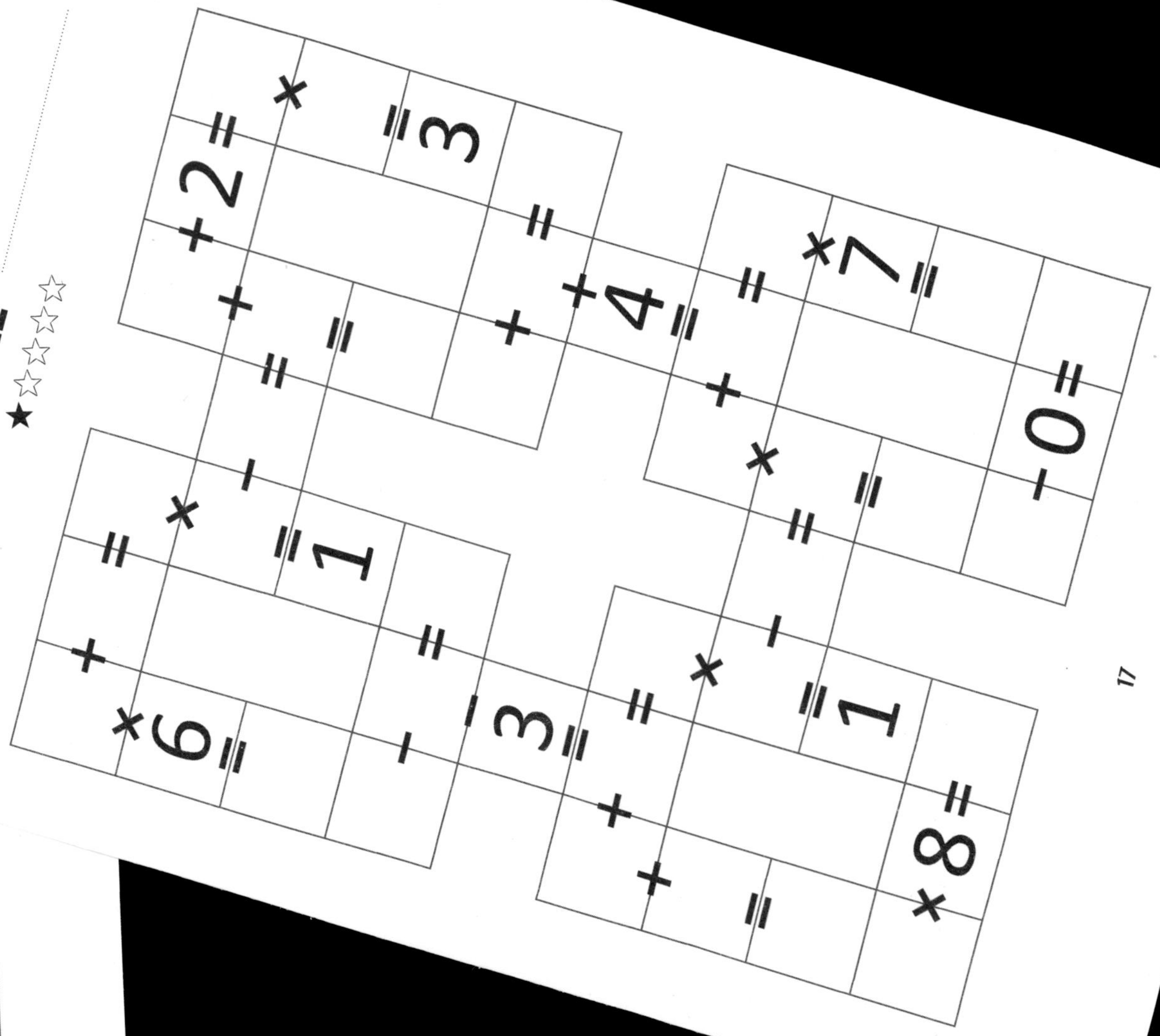

015

★☆☆☆☆

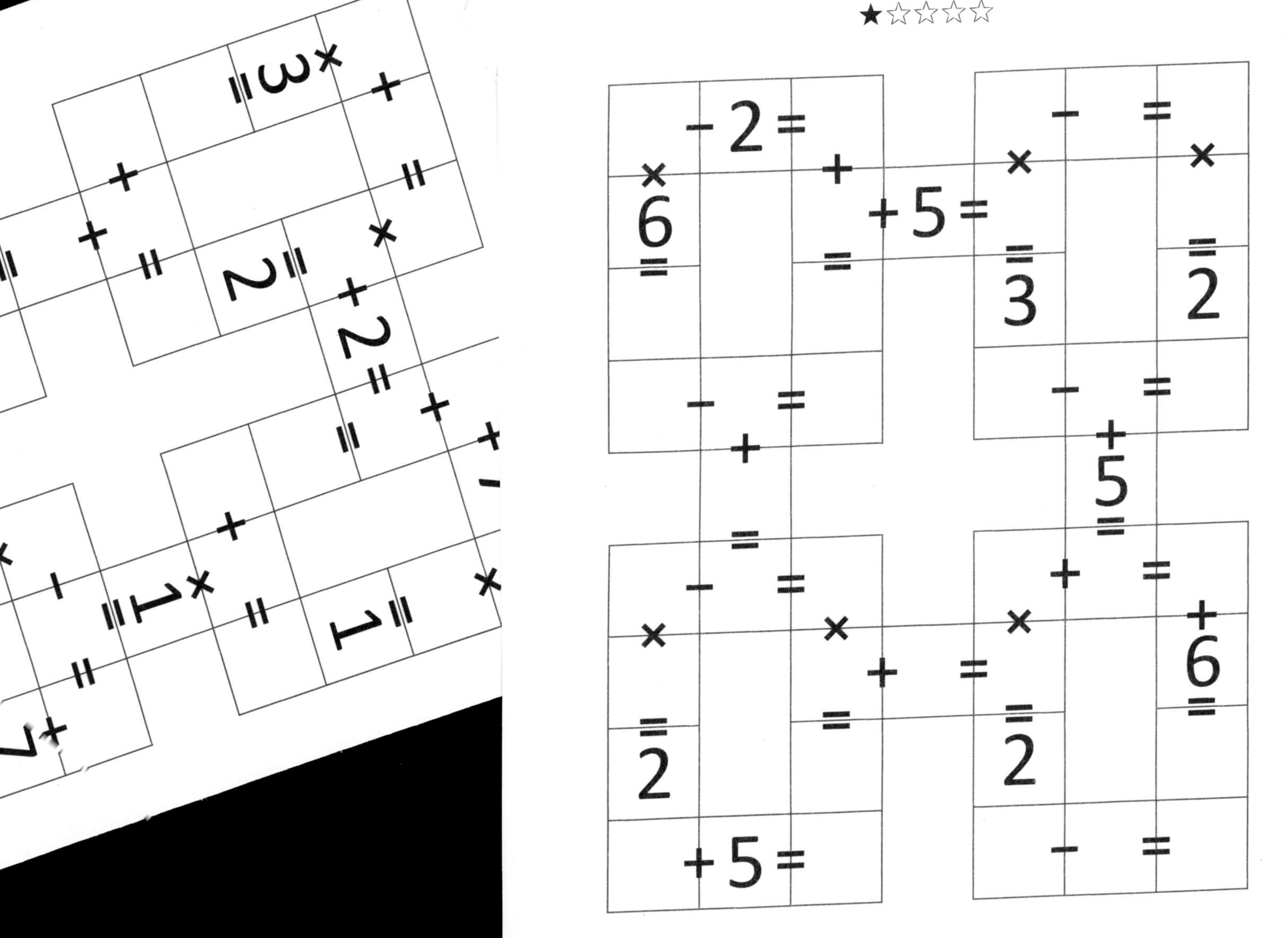

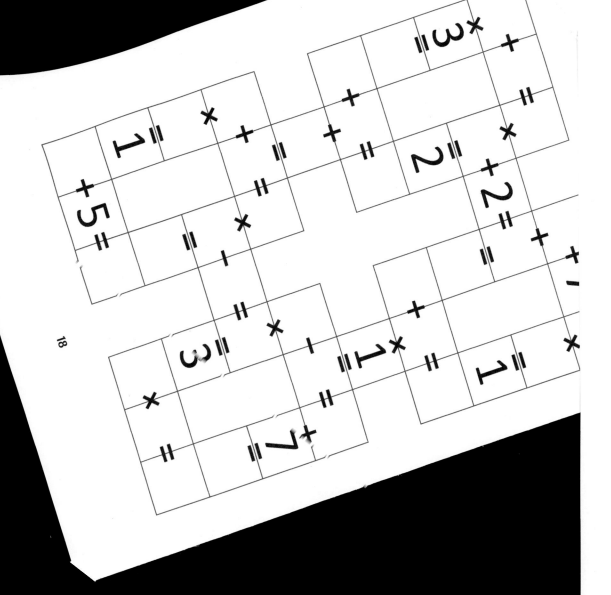

016

★☆☆☆☆

−6=
+ × + 2 = × ×
3 1 1
= = = =
+ = + =
− +
= =
+ = + =
+ + + ×
7 −5= 3
= = = =
×8= +5=

017

★☆☆☆☆

\+ = +1=

× × + ×

4 −3= 9

= = = =

1

\+ = − =

− +

= =

\+ = + =

\+ × + ×

+1=

= = = =

1 4

−1= −3=

018

★☆☆☆☆

+ = − 4 =

× × × +

2 × 1 = 8

= = = =

1

× = + =

+ +

= =

+ = − =

× × × ×

4 × 1 =

= = = =

4

− 4 = − 7 =

019

★☆☆☆☆

− 7 =

× + − 2 = × − = +

7

= = = =

2 1

− = − =

− −

= =

+ = + =

+ × − 1 = × ×

= = = =

1 3

− 1 = − 4 =

020

★☆☆☆☆

+ 3 = − 0 =

+ × + +

− 2 = 5

= = = =

1

+ = × =

+ +

2

= =

+ = − =

× × × ×

8 − =

= = = =

2

− 5 = − 6 =

021

★☆☆☆☆

+ 3 =　　+ 0 =

+　×　− 4 =　+　×

=　=　=　=

2　1

+ =　× − =

+　−

=　=

+ =　+ =

×　×　− 5 =　×　×

9

=　=　=　=

1

+ 5 =　+ 4 =

022

★☆☆☆☆

− =
+ ×
5 × 1 = + + 5 = ×
= = = =
1 3
− = × =
+ −
5
= =
− = − =
× + × +
4 − = 8
= = = =
− 1 = + 6 =

023

★☆☆☆☆

+ 3 =

× ×

− 2 =

× ×

= = = =

3 3 1

− = − =

+ −

= =

− = + =

× + + +

+ 4 =

9

= = = =

4

+ 1 = × 5 =

024

★☆☆☆☆

	−5=	
+5=		×=
	−=	

+4=

	+1=	
×=2		+=
	+=	

−=

+4=

	−=	
×=		×=4
	×9=	

+=

	−=	
×=3		×=
	−2=	

025

★☆☆☆☆

− 2 =   − =

+   ×   ×   +

7   + =   8

=   =   =   =

2

− =   + =

×   −

2   4

=   =

+ =   − =

×   +   ×   +

2   + =   9

=   =   =   =

1

+ 2 =   × =

026

★☆☆☆☆

+ = +7=
× × + ×
+1=
= = = =
3 2 1
× = + =
+ −
= =
+ = − =
× + × ×
+4=
= = = =
3 2
+0= −7=

027

★☆☆☆☆

+ 0 =   − =

× + × ×

3 + 4 = 3

= = = =

3

+ =   − =

− −

= =

+ =   + =

+ × + ×

8 − 4 =

= = = =

2   3

− =   × 6 =

028

★☆☆☆☆

− 6 = − 6 =

+ × + ×

9 + 3 = 6

= = = =

× = − =

− +

= =

+ = − =

+ × + ×

5 + 1 = 3

= = = =

2

+ = − 4 =

029

★☆☆☆☆

− 4 =   − =

× +   × ×

+ 3 =

= =   = =

3   1 1

− =   + =

−   −

=   =

− =   − =

+ +   × ×

5   + 4 =

= =   = =

4

× 3 =   − 4 =

030

★☆☆☆☆

− 6 =				−	=		
×	+	+	=	×	×		
=	=			=	=		
4				4	3		
× =				− =			
+ 4 =				− 4 =			
+ =				− =			
×	×	+	=	×	+ 2 =		
=	= 3			= 4			
− 5 =				− =			

031

★★☆☆☆

− 2 =

× × +4 = × +

4

= = = =

1

× =

+ =

−

2

=

−

2

=

− =

+ =

× + +4 = + ×

= = = =

3

× 4 =

× 6 =

032

★★☆☆☆

− = + =

× + −1= + ×

3

= = = =

2

+ = − =

− −

4 4

= =

− = − =

× + −4= + ×

7

= = = =

3

+3= −1=

033

034

★★☆☆☆

− 1 =   − =

× + × +

8 + 1 =

= = = =

1

+ =   + =

+ +

2   4

= =

− =   + =

× × × +

+ 2 =   4

= = = =

× 9 =   − 7 =

035

★★☆☆☆

+ 4 =   + =

× + + ×

+ 4 =

= = = =

3   1

− =   − =

+   −

4   4

=   =

+ =   − =

+ × × +

7 + 4 =

= = = =

+ 6 =   + 5 =

036

+ = − =

+ × × +

2 +4=

= = = =

3

+ = + =

+ +

3 4

= =

− = + =

× + + ×

−2=

= = = =

2 4

×1= ×2=

037

★★☆☆☆

− 3 =　　　　+ 5 =

×　　+　　　　+　　×

− 1 =

=　　=　　　　=　　=

4

+ =　　　　× =

−　　　　+

2　　　　4

=　　　　=

− =　　　　+ =

×　　×　　　　×　　×

− 4 =　　　　5

=　　=　　　　=　　=

× 9 =　　　　− 0 =

038

★★☆☆☆

− 5 = + 5 =

× × × ×

− 3 =

= = = =

1

− = × =

− +

4 4

= =

− = + =

+ × + ×

7 − 1 =

= = = =

2

+ 2 = − =

039

★★☆☆☆

	−		=	
×				+
4				
=				=
	+		=	

+3=

	−		=	
×				+
=				=
2				
	+		=	

+ 3 = (left) ; + 3 = (right)

	+		=	
×				+
=				=
2				
	×	1	=	

−4=

	−		=	
+				×
				3
=				=
	−	0	=	

040

★★☆☆☆

+ 2 =   + =

×   +   +   ×

− 4 =

=   =   =   =

3   2

× =   − =

−   +

2   4

=   =

+ =   + =

+   ×   ×   +

− 4 =   7

=   =   =   =

1   1

× =   × =

041

★★☆☆☆

− 6 =
× +
− 2 =
× +
= =
= =
1

− =
+ =
+ −
4 1
= =

+ =
+ =
× + × ×
9 − 4 = 8
= = = =

× 1 =
× 4 =

042

★★☆☆☆

− 4 =   − =

× +   × +

+ 1 =   2

= =   = =

4

+ =   − =

+   +

1   4

=   =

+ =   − =

+ ×   × +

− 4 =   6

= =   = =

4

− =   − 4 =

043

★★☆☆☆

− 1 =

+ × × +

6 + 3 =

= = = =

2

+ = + =

− −

1 3

= =

+ = + =

× + + ×

+ 3 =

= = = =

3

− 4 = × 9 =

044

★★☆☆☆

− 3 = − =

× + + 1 = × +

= = = =

1 1

− = + =

− −

2 1

= =

− = − =

+ × + +

9 × 1 = 2

= = = =

× 1 = × =

045

★★☆☆☆

+ =   + 5 =

+ ×   × ×

− 4 =

= = 4   = = 2

− =   + =

+ 3 =   + 3 =

+ =   + =

× +   × ×

7 = = × 1 = = 4 =

× 2 =   × =

046

★★☆☆☆

+ 3 =　　　　− 5 =

×　　+　　　　×　　+

4　　− 4 =

=　　=　　　　=　　=

+ =　　　　+ =

+　　　　−

2　　　　2

=　　　　=

+ =　　　　− =

+　　×　　　　+　　×

− 3 =

=　　=　　　　=　　=

2　　　　4

× =　　　　+ 7 =

047

★★☆☆☆

+ 4 =   + 2 =

× 9 =   + − 3 =   + =   × = 2

− =   × =

− 2 =   + 3 =

− =   − =

× × − 2 = × +

= =   = 4 =

× 7 =   + =

048

★★☆☆☆

− 2 =   − 1 =

×   +   +   ×

+ 3 =   8

=   =   =   =

1

− =   − =

+   +

4   4

=   =

+ =   + =

×   ×   ×   ×

× 1 =

=   =   =   =

× 6 =   × 9 =

049

★★☆☆☆

+ = − 3 =

+ × × +

+ 3 =

= = = =

1

− = + =

−

4

=

−

3

=

+ = + =

× + + ×

− 1 =

= = = =

2 1

+ 5 = × 4 =

050

★★☆☆☆

+ =
+ ×
2 +3=
= =

× =
+
4
=
+ =
× ×
= -1=
=
-7=

-5=
+ ×
+3=
= =
1
× =
+
2
=
- =
× ×
-1=
= =
2
+1=

051

★★☆☆☆

052

★★☆☆☆

	+ 3 =					+	=	
+		×			×			×
			+ 1 =					9
=		=			=			=
	−	=				×	=	
	−					−		
	4					3		
	=					=		
	+	=				+	=	
+		×			+			×
7			− 3 =					
=		=			=			=
								4
	+ 2 =					− 3 =		

053

★★☆☆☆

\+ = +3 =

\+ × × ×

3 −4 =

= = = = =

4

× = − =

− −

4 4

= =

\+ = − =

× + + ×

+3 =

= = = =

3

+3 = −7 =

054

★★☆☆☆

	+		=						−		=	
+				×				×				+
2					−	1	=					6
=				=				=				=
	+		=						+		=	
		×								−		
		1								4		
		=								=		
	−		=						+		=	
×				+				+				×
					+	2	=					
=				=				=				=
4												3
	−	3	=						+	3	=	

055

★★☆☆☆

\+ = − 2 =

\+ × + ×

− 4 = 9

= = = =

1

− = + =

× +

1 4

= =

− = + =

× + × ×

8 − 4 =

= = = =

− 1 = + 7 =

GARAM

056

★★☆☆☆

− 2 =   − 0 =

×   ×   ×   ×

× 1 =   8

=   =   =   =

4

× =   − =

−   +

3   1

=   =

− =   + =

×   +   ×   ×

4   + 4 =

=   =   =   =

+ =   × 5 =

057

★★☆☆☆

− 6 =

× ×

+ 1 =

× +

= = = =

1

× =

+ =

+ −

4 1

= =

− =

+ =

× + × +

3 − 1 = 7

= = = =

− 1 =

− 2 =

058

★★☆☆☆

+ 3 =　　+ 1 =

×　　+　　+　　×

7　　− 1 =

=　　=　　=　　=

3

+ =　　× =

+　　−

1　　4

=　　=

+ =　　+ =

+　　×　　×　　×

− 4 =

=　　=　　=　　=

1

× =　　− 7 =

059

★★☆☆☆

	+		=	
+				×
2				
=				=
	×		=	

+ 2 =

	+	5	=	
×				×
=				=
				3
	−		=	

+ 4 =

× 1 =

	+		=	
×				×
=				=
	×	6	=	

+ 1 =

	−		=	
+				×
				9
=				=
	−	4	=	

060

★★☆☆☆

+ 7 =　　　+ 5 =

+　×　　　+　×

× 1 =

=　=　　　=　=

3

+　=　　　+　=

−　　　　+

4　　　　2

=　　　　=

+　=　　　−　=

×　×　　　×　×

× 1 =　　　2

=　=　　　=　=

3

+ 1 =　　　+　=

061

★★★☆☆

062

− 1 =

× +

+ 4 =

× +

= = = =

2 2

× = + =

+ −

2 1

= =

− = + =

+ × + ×

8 − 2 =

= = = =

2

+ 5 = − =

063

+ =

+ ×

− 1 =

× +

= = = =

4 1

− =

+ =

− −

4 1

= =

− =

+ =

× +

× +

+ 3 =

7

= = = =

3

− 2 =

− 2 =

064

★★★☆☆

− 0 =

+ 3 = × − 3 = × = 1 − = + =

− = + 3 = + − 3 =

− = × 3 = + − 2 = × = + 9 =

+ = + 2 =

065

★★★☆☆

\+ 2 = − 7 =

× × × +

8 + 2 =

= = = =

× = × =

− +

4 3

= =

\+ = + =

× × + ×

\+ 4 = 3

= = = =

\+ 7 = + 2 =

066

− = − 1 =

× + × ×

+ 2 = 3

= = = =

2

+ = − =

+ +

1 4

= =

− = + =

× × + ×

+ 3 = 4

= = = =

× 5 = − 1 =

067

★★★☆☆

068

★★★☆☆

+ 0 = ×

+ 7 =

+ 4 =

+ 5 = ×

+ =

× =

× =

+ 4 =

− 4 =

+ =

− =

× ×

− 2 =

+ ×

= =

= = 2

× 8 =

− 3 =

069

★★★☆☆

−6 =   +1 =

+   ×   +   ×

−3 =

=   =   =   =

1

× =   × =

+   +

2   4

=   =

− =   + =

×   ×   +   ×

+1 =

=   =   =   =

2

×6 =   +2 =

070

★★★☆☆

−3=
× + +3= + +5= ×
= = = = 4
+= +=
+ +
4 4
= =
−= +=
+ × × ×
9 ×1=
= = = =
3
+3= ×=

071

★★★☆☆

+ 5 =

× 5 = × + 4 = + × 7 =

− =

+ 1 = − 4 =

− = − =

× = × − 2 = × = + 2 =

= 1

× 3 = × =

072

★★★☆☆

− 0 =　　　　− =

×　　+　　×　　+

9　　+ 3 =　　2

=　　=　　=　　=

+ =　　　　− =

+　　　　−

3　　　　4

=　　　　=

+ =　　　　− =

×　　×　　×　　+

− 2 =

=　　=　　=　　=

4　　　　1

− 4 =　　　　+ =

073

★★★☆☆

\+ 7 =   − 0 =

\+   ×   ×   +

− 4 =   7

=   =   =   =

\+ =   × =

−   −

4   4

=   =

\+ =   − =

×   ×   ×   ×

\+ 4 =

=   =   =   =

2

\+ 7 =   − 0 =

074

★★★☆☆

− 1 =   − =

×   +   ×   +

− 4 =   2

=   =   =   =

2

× =   − =

×   −

1   1

=   =

\+ =   − =

×   +   +   +

7   + 3 =   3

=   =   =   =

× 2 =   × =

075

★★★☆☆

076

★★★☆☆

	+ 1 =	
+ 4 =		× =

− 1 =

	+ =	
+ =		× 4 =

	− =	

+ 4 =

	− =	

− 4 =

	− =	
× 4 =		+ =

− 4 =

	− =	
+ =		× = 1

	+ =	

	+ 1 =	

077

− 5 = + =

× × × +

6 + 4 = 4

= = = =

+ = + =

+ +

3 3

= =

+ = − =

+ × × ×

+ 3 = 9

= = = =

2

− = − 1 =

078

★★★☆☆

+ 3 =   + 2 =

+   ×   ×   +

+ 4 =   5

=   =   =   =

− =   × =

−   +

4   4

=   =

+ =   + =

+   ×   +   ×

− 1 =   8

=   =   =   =

− 5 =   × 1 =

079

★★★☆☆

	− 7	=						−		=			
×			+				×			+			
				× 1	=					2			
=			=				=			=			
	−		=					−		=			
	+							−					
	2							4					
	=							=					
	−		=					−		=			
×			×				+			×			
7				+ 2	=								
=			=				=			=			
										3			
	+ 1	=						× 2	=				

080

★★★☆☆

−1=
×
×
+4=
×
−
=
+
=
=
2
=
=
3
×
=
+
=
−
+
4
2
=
=
+
=
+
=
×
+
×
×
−3=
=
=
=
=
1
+2=
+7=

081

★★★☆☆

\+ 6 =   − 7 =

\+   ×   − 3 =   ×   +

=   =   =   =

× =   − =

\+   −

2   1

=   =

− =   + =

\+   ×   ×   +

6   × 1 =   7

=   =   =   =

\+ 6 =   − 7 =

082

★★★☆☆

− 1 =   + 4 =

\+ ×   + ×

8   − 1 =

= =   = =

− =   − =

\+   +

3   3

=   =

\+ =   + =

× ×   × ×

− 2 =

= =   = =

3

× 5 =   × 6 =

083

★★★☆☆

+0=

× + × +

4 +4=

= = = =

4

+ = + =

+ −

4 4

= =

+ = − =

× × + ×

−1= 7

= = = =

+8= +4=

084

★★★☆☆

+ 3 =　　　　− 5 =

×　　×　　+　　×

− 3 =　　8

=　　=　　=　　=

3

× =　　　　× =

+　　　　+

4　　　　2

=　　　　=

− =　　　　− =

×　　+　　×　　+

− 2 =　　6

=　　=　　=　　=

× 3 =　　　　+ =

085

★★★☆☆

− 2 = − =

× + × +

+ 3 =

= = = =

2 2

× = + =

+ −

2 1

= =

− = − =

× × × ×

8 + 2 =

= = = =

+ 3 = × 5 =

086

★★★☆☆

+ = − =
+ × × +
2 +2= 6
= = = =
+ = + =
+ −
4 3
= =
− = + =
× + × ×
9 ×1=
= = = =
4
+1= −0=

087

★★★☆☆

− = × + 2 − 4 = − 3 = × + = = = =

+ = + = − 4 = − 4 =

+ = − = × × + × + 4 = = = = =

4 1

− 2 = × 6 =

088

★★★☆☆

+ 2 =　　　− 5 =

+　　×　− 3 =　+　　×

6　　　　　　　　8

=　　=　　=　　=

− =　　　　× =

−　　　　　+

1　　　　　4

=　　　　　=

− =　　　　+ =

×　　+　+ 1 =　+　　×

　　　　　　　　7

=　　=　　=　　=

× 4 =　　　　− =

089

★★★☆☆

+ 3 =

− =

× + × +

− 2 =

= = = =

2 1

× = + =

− +

2 3

= =

− = − =

× + × ×

8 + 4 =

= = = =

3

+ = + 1 =

090

★★★☆☆

− =
× +
−3=
= =
1
+ =
−
4
=
− =
× +
3
+4=
= =
+ =

−4=
+ ×
4
= =
− =
+
1
=
+ =
× +
5
= =
−0=

091

★★★★☆

- 3 = - 4 =

× 9 = × + 4 = + = × = 2

× = × =

+ 3 = − 1 =

+ = + =

× = × = 2 + 4 = + = × 9 =

- 6 = - 5 =

092

− = − 1 =

× × + ×

− 4 =

= = = =

4 2 3

− = × =

− −

2 2

= =

− = − =

× + × +

2 + 2 =

= = = =

2

− 5 = + 2 =

093

★★★★☆

+3=

×

7

=

+

=

+1=

+4=

×

=

3

+

=

+ =

× =

−

2

=

−

3

=

− =

− =

×

=

3

×

=

−3=

×

=

2

+

=

−1=

−4=

094

★★★★☆

095

★★★★☆

− 1 =   − 1 =

×   +   ×   ×

− 1 =

=   =   =   =

2   2

+   =   −   =

−   +

3   2

=   =

−   =   −   =

×   +   ×   +

+ 3 =

=   =   =   =

3   4

× 1 =   − 4 =

096

★★★★☆

+ 0 =　　　　+ 2 =

+　　×　　　　+　　+

− 1 =　　5

=　　=　　　　=　　=

1

× =　　　　× =

+　　　　×

2　　　　1

=　　　　=

− =　　　　+ =

×　　+　　　　×　　+

9　　− 1 =

=　　=　　　　=　　=

3

+ 2 =　　　　+ 6 =

097

★★★★☆

− 6 =   + 1 =

×   +   ×   ×

× 1 =

=   =   =   =

1   1

− =   × =

−   −

2   3

=   =

− =   − =

+   ×   +   ×

9   + 2 =

=   =   =   =

1

+ 3 =   + 4 =

098

+ =   +6=

×   ×   ×   +

×1=   6

=   =   =   =

1   1

+ =   × =

−   −

1   2

=   =

− =   + =

×   +   +   ×

3   +1=   8

=   =   =   =

+2=   +2=

099

\+ = +4 =

× × + ×

2 +1 =

= = = =

3 4

− = + =

− −

1 2

= =

\+ = + =

× + × ×

7 −3 =

= = = =

2

×1 = ×4 =

100

★★★★☆

−0=
+ ×
×1=
× × 4
= = 2 = 4 =
+ =
× =
+ −
3 3
= =
− =
− =
+ × + ×
+1=
= = 4 = = 4
−1=
−6=

101

− 4 =   − 1 =

\+   ×   +   ×

− 4 =

=   =   =   =

1   2

\+ =   × =

×   −

1   1

=   =

\+ =   + =

×   ×   ×   ×

− 2 =

2

=   =   =   =

1

\+ 1 =   − 1 =

★★★★☆

− =		− 2 =
+ ×		× ×
3	+ 4 =	
= =		= =
1		1
+ =		× =
+		−
3		4
=		=
− =		+ =
× +		+ ×
7	× 1 =	
= =		= =
		4
+ 0 =		× 2 =

103

+ 4 =   − 6 =

×   ×   − 1 =   +   +

9

=   =   =   =

2

− =   × =

−   +

2   4

=   =

+ =   − =

+   ×   ×   ×

8   − 1 =   6

=   =   =   =

3

+ 1 =   − =

104

★★★★☆

	+		=	
+				×
7				
=				=
				3
	−		=	

+2=

	+3=	
+		×
		7
=		=
	×	=

+ 3 =

− 2 =

	+		=	
+				×
7				
=				=
	+1=	

+3=

	−		=	
×				×
=				=
3				
	+3=	

105

+ 4 = + 2 =

× + × ×

+ 4 =

= = = =

1 1

− = × =

+ −

3 4

= =

− = + =

+ × × +

5 + 3 = 9

= = = =

1 1

+ = + =

106

− 5 = − 0 =

+ + + ×

3 − 2 =

= = = =

2

× = + =

+ +

3 3

= =

+ = + =

× × × ×

+ 4 =

= = = =

2 3

− 6 = − 2 =

107

− 5 =

+ ×

5

+ 4 =

× + 4 = ×

2

= = = =

− = + =

+ +

3 1

= =

+ = + =

× × × +

+ 3 =

5

= = = =

1 1 1

− = × =

108

+ 0 =　　　　− 3 =

+　　×　　×　　×

+ 1 =　　8

=　　=　　=　　=

2

× =　　− =

−　　+

4　　4

=　　=

− =　　+ =

+　　×　　×　　+

− 3 =　　8

=　　=　　=　　=

3

− 0 =　　+ 1 =

109

+ 6 =   + =

\+ × × ×

9 − 1 =

= = = =

1   2

× =   × =

\+   −

2   3

=   =

− =   + =

\+ × + ×

9 − 3 =

= = = =

1   2

× =   + 3 =

110

★★★★☆

+ 1 =　　　− =

× ×　　× ×

− 3 =

6

= =　　= =

4　　1

× =　　× =

−　　+

3　　1

=　　=

+ =　　+ =

× ×　　× ×

− 1 =

= =　　= =

3　　3

− 6 =　　− 4 =

111

★★★★☆

	−	2	=	
+				×
=				= 4
	+		=	
		− 3 =		
	−		=	
× 4 =				+
	−	4	=	

× 1 =   + 2 =

	+	7	=	
+				×
=				= 2
	+		=	
		+ 1 =		
	−		=	
+				× 5 =
	−	1	=	

112

★★★★☆

− 5 =   − =

×   ×   ×   ×

+ 1 =

=   =   =   =

2   3   3

× =   × =

−   ×

1   1

=   =

+ =   + =

×   ×   ×   ×

4   − 2 =   4

=   =   =   =

3

− =   + 5 =

113

★★★★☆

+0=
+ ×
6 −2= + × 4
= = = =
− = × =
− −
4 3
= =
+ = + =
+ × × +
6 −1= 7
= = = =
2
×4= − =

114

★★★★☆

+2=
×
×
6
=
=
−3=
−4=
+
+
7
=
=
×=
×=
−
×
3
1
=
=
+=
+=
+
×
×
+
7
+4=
=
=
=
=
2
+0=
+1=

115

★★★★☆

− 0 =　　　　　　+ 6 =

×　　　+　　　　+　　　×

× 1 =　　　　6

=　　　=　　　=　　　=

4

+　=　　　　+　=

−　　　　　　　+

4　　　　　　　4

=　　　　　　　=

+　=　　　　+　=

+　　　×　　　+　　　×

5　　　+ 3 =

=　　　=　　　=　　　=

4　　　　　　　2

−　=　　　　+ 7 =

116

★★★★☆

− 2 =

× × + 4 = × + 4 = +

= 3 = = 3 =

× = − =

− 3 + 3

= =

+ = − =

+ × + 1 = × ×

= = 1 = 3 =

− 1 = − 0 =

117

+ 4 =　　　　− 3 =

×　　+　+ 3 =　×　　+

3　　　　　　　　　　7

=　　=　　　　=　　=

− =　　　　+ =

+　　　　　−

2　　　　　1

=　　　　　=

− =　　　　− =

×　　×　− 2 =　×　　+

8

=　　=　　　　=　　=

4　　2

+ =　　　　× 6 =

118

− 3 =   + 1 =

×   +   +   ×

+ 2 =   4

=   =   =   =

4

− =   × =

+   −

3   4

=   =

− =   − =

×   +   ×   ×

× 1 =   6

=   =   =   =

4

× 1 =   × 2 =

119

	−		=	
×		×		
3				
=		=		
		3		
	+		=	
		−		
		1		
		=		

+2=

	−		=	
×		×		
=		=		
2		2		
	×		=	
		×		
		1		
		=		

	+		=	
×		×		
9				
=		=		
	+3=			

+4=

	−		=	
×		+		
		6		
=		=		
	−4=			

GARAM

120

\+ 1 =   + =

× ×   × +

\+ 1 =   9

= =   = =

3   4

− =   − =

\+   −

1   2

=   =

− =   + =

× ×   × +

4   − 1 =   6

= =   = =

4   3

\+ =   × =

121

★★★★★

+ = × + 2 = × − 5 = +
2 = = = =
× = + = + 4 = + 1 =
− = + = × + 2 = × − 4 = × 3 = × = =
− 1 = + 0 =

122

+ 0 =   − =

×   ×   +   +

3   − 1 =   2

=   =   =   =

× =   × =

−   −

4   1

=   =

+ =   − =

×   ×   +   ×

+ 2 =   7

=   =   =   =

− 7 =   × 3 =

★★★★★

− =
+ × − 3 =
2 − 4 = + × 7
= = = =
− = × =
+ −
4 3
= =
+ = − =
× + × +
+ 2 = 5
= = = =
2
− 2 = + =

124

+ =
+ ×
3 + 3 = + 4
= = = =
× = × =
– –
2 4
= =
+ = – =
+ × + ×
+ 2 = 4
= = = =
1
× = – 6 =

+ 1 =

125

\+ = +3 =

\+ × × ×

6 −2 =

= = = =

4

− = − =

− +

4 3

= =

\+ = − =

× × × ×

9 −1 =

= = = =

2

+1 = × =

126

★★★★★

\+ = − 4 =

\+ + + ×

2 + 1 = 6

= = = =

× = × =

− +

1 4

= =

− = + =

× + × ×

8 + 2 =

= = = =

2

\+ = − 1 =

127

− 6 = − 0 =

× + + ×

− 2 = 4

= = = =

× = × =

− −

2 4

= =

+ = − =

× + × ×

4 × 1 =

= = = =

+ 4 = − 8 =

128

\+ = + =

\+ × × ×

5 + 1 = 7

= = = =

− = × =

− −

2 4

= =

\+ = − =

\+ × × +

8 × 1 = 9

= = = =

× 3 = + 4 =

129

★★★★★

+7=
+ +
= =
+2=
+ ×
= =
× =
+
2
=
− =
+
2
=
+ =
× +
3
= =
+3=
+ =
× ×
= =
2
+2=
×2=

130

★★★★★

+6= +6=

+ × × +

−2= 9

= = = =

× = + =

+ +

3 1

= =

+ = − =

+ × + ×

+3=

= = = =

1 3

× = +0=

131

\+ 6 =   − =

\+ × × +

\+ 4 =

= = = =

4

× =   + =

\+   −

1   4

=   =

\+ =   + =

× + × +

\+ 2 =   4

= = = =

2

− 2 =   + 1 =

132

★★★★★

□ + 1 = □
× ×
9 □ + 2 = □
= = = =
□ □ 4 □
□ × □ = □   □ + □ = □
+ −
4 3
= =
□ + □ = □   □ + □ = □
× × × +
□ □ + 3 = □ 5
= = = =
□ − 7 = □   □ − 3 = □

133

+ 3 =　　　　　　　　− 4 =

\+　　　×　　　　　　×　　　×

\+ 4 =

=　　　=　　　　　　=　　　=

　　　　　　　　　　　　　2

− =　　　　　　　　− =

−　　　　　　　　　　+

3　　　　　　　　　　4

=　　　　　　　　　　=

− =　　　　　　　　− =

×　　　+　　　　　　×　　　+

× 1 =

=　　　=　　　　　　=　　　=

1　　　　　　　　　3

− 6 =　　　　　　　　+ =

134

− 1 = + 3 =

× × + ×

− 2 =

= = = =

2 1

− = × =

− +

3 2

= =

+ = − =

× × + ×

× 1 =

= = = =

× 8 = − 7 =

135

★★★★★

+ 2 = + 3 =

+ × + ×

9 + 3 =

= = = =

× = − =

− −

3 3

= =

+ = − =

× × × +

− 4 = 4

= = = =

4

× 1 = + =

★★★★★

− 0 = + =
+ × + ×
8 − 4 =
= = = =
2
+ = − =
+ +
3 3
= =
+ = − =
× + × ×
− 1 =
= = = =
4 2
× 1 = × =

137

★★★★★

138

★★★★★

+7=

+ ×

+4=

× × 7

= = = =

+ =   − =

× +

1   4

= =

− =   − =

× × + ×

+4=   4

= = = =

×8=   −2=

139

★★★★★

+6= −0=

\+ × × +

+3= 3

= = = =

× = + =

\+ −

2 1

= =

\+ = − =

\+ × + ×

+4=

= = = =

1 1

× = ×6=

★★★★★

+ 1 =　　　　− 3 =

× 　 + 　　　　× 　 +

9 　 − 3 = 　　 6

= 　 = 　　　　= 　 =

+ 　 = 　　　　+ 　 =

+ 　　　　　　　 ×

4 　　　　　　　 1

= 　　　　　　　 =

+ 　 = 　　　　− 　 =

× 　 + 　　　　× 　 +

− 2 = 　　　　 2

= 　 = 　　　　= 　 =

× 5 = 　　　　− 　 =

141

− 3 =　　　　− 7 =

×　　+　　×　　+

9　　− 4 =

=　　=　　=　　=

× =　　− =

+　　−

1　　2

=　　=

− =　　− =

×　　+　　×　　×

− 2 =

=　　=　　=　　=

4

× 6 =　　+ 7 =

142

★★★★★

+ 3 =
− 3 =
+ ×
− 2 =
+ ×
= =
= =
1
− =
− =
− −
4 1
= =
− =
− =
+ ×
+ 4 =
+ ×
= =
= =
4
− 7 =
− 6 =

143

★★★★★

− 3 =   + =

×   +   +   ×

\+ 2 =

=   =   =   =

4   2

\+ =   − =

\+   −

4   4

=   =

\+ =   − =

\+   ×   ×   +

\+ 3 =   3

=   =   =   =

3

− =   − 1 =

144

★★★★★

− 5 =   + =

×   ×   +   ×

+ 4 =   2

=   =   =   =

× =   − =

−   −

4   1

=   =

− =   + =

×   ×   ×   ×

+ 4 =

=   =   =   =

2   1   3

− 0 =   × =

145

★★★★★

+ =
+ × + 2 = × +0= +
3
= = = =
2
− = − =
− +
4 1
= =
+ = − =
× + × 1 = + ×
= = = =
2
+7= −0=

GARAM

146

★★★★★

− = +4 =

× + + ×

+3 =

= = = =

1

+ = − =

+ −

3 1

= =

− = + =

× × + ×

−4 =

= = = =

3 3

×4 = +2 =

147

★★★★★

+ 4 = − 7 =

× + × +

− 1 =

= = = =

1

+ = − =

− +

4 3

= =

+ = − =

+ × × +

4 − 4 =

= = = =

× 1 = + 3 =

148

★★★★★

−1=

×

9

=

+

+3=

×

−2=

×

=

=

2

+ =

× =

+

3

=

−

3

=

+ =

− =

×

+

−2=

×

×

= = = =

×7=

×9=

149

★★★★★

+ 7 =
+ + + 1 = + − 4 = ×
8
= = = =
× = × =
+
1
=
+ = − =
+ × + 4 = + ×
= = = =
3
− 3 = − 4 =

150

− 0 =   − =

× + − 4 = × +

8   5

= = = =

− =   + =

−   −

4   2

=   =

+ =   − =

× + + 1 = × +

= = = =

3   1

− 4 =   + =

001

```
2+1=3   9-2=7
+   ×   +   ×
9   6-5=1   5
=   =   =   =
1   1   1   3
1×8=8   0+5=5
  -       +
  4       1
  =       =
9-4=5   2+6=8
×   ×   ×   ×
9   7+1=8   7
=   =   =   =
8   3   1   5
1×5=5   6+0=6
```

002

```
8-6=2   8+1=9
×   ×   ×   +
3   5-1=4   7
=   =   =   =
2   1   3   1
4×0=0   2×3=6
  +       -
  5       1
  =       =
4+5=9   7+2=9
×   +   +   ×
4   6-2=4   3
=   =   =   =
1   1   1   2
6-1=5   1+6=7
```

003

```
3+2=5   3+6=9
+   +   ×   +
7   5-1=4   5
=   =   =   =
1   1   1   1
0×6=0   2×2=4
  -       +
  5       1
  =       =
7+1=8   8-3=5
+   +   ×   +
3   2+3=5   5
=   =   =   =
1   1   4   1
0×3=0   0×4=0
```

004

```
9-6=3   2+2=4
+   ×   ×   +
3   4+2=6   8
=   =   =   =
1   1   1   1
2×1=2   2+0=2
  -       +
  1       2
  =       =
4+0=4   8-2=6
+   ×   ×   +
6   7+1=8   4
=   =   =   =
1   2   6   1
0+8=8   4-4=0
```

005

```
3+2=5   3+6=9
+   ×   +   +
9   9-1=8   5
=   =   =   =
1   4   1   1
2+3=5   1×4=4
  +       ×
  3       1
  =       =
8-6=2   3+4=7
+   ×   ×   +
3   6+1=7   9
=   =   =   =
1   1   2   1
1+1=2   1+5=6
```

006

```
3+5=8   4+4=8
×   +   +   ×
5   7-1=6   5
=   =   =   =
1   1   1   4
5×1=5   0×6=0
  -       -
  1       4
  =       =
5+0=5   5+2=7
×   ×   ×   +
5   9×1=9   8
=   =   =   =
2   4   4   1
5×1=5   5×1=5
```

007

```
9-1=8   9-3=6
×   ×   +   ×
4   4-1=3   2
=   =   =   =
3   3   1   1
6-4=2   2+0=2
  -       +
  2       3
  =       =
6-2=4   7-3=4
+   ×   ×   ×
4   5+2=7   8
=   =   =   =
1   2   4   3
0×9=0   9-7=2
```

008

```
9-1=8   8+0=8
+   ×   ×   +
8   5-1=4   3
=   =   =   =
1   4   3   1
7-7=0   2-1=1
  -       +
  1       3
  =       =
3+6=9   8-4=4
×   +   +   ×
7   2×2=4   4
=   =   =   =
2   1   1   1
1-0=1   2+4=6
```

009

```
2-0=2   3+5=8
+   ×   +   +
9   8+1=9   8
=   =   =   =
1   1   1   1
1×6=6   2×3=6
  -       +
  3       1
  =       =
6+3=9   8-4=4
+   +   +   +
4   1+1=2   6
=   =   =   =
1   1   1   1
0×2=0   0×2=0
```

010

```
9-6=3   4-0=4
×   ×   ×   ×
3   6+1=7   8
=   =   =   =
2   1   2   3
7+1=8   8-6=2
  -       -
  1       5
  =       =
8-0=8   3-1=2
+   ×   ×   +
3   5+2=7   8
=   =   =   =
1   4   2   1
1-1=0   1-1=0
```

011

```
5+2=7   7+2=9
×   +   +   +
9   8-3=5   9
=   =   =   =
4   1   1   1
5+0=5   2×4=8
  +       +
  1       2
  =       =
3+1=4   3+6=9
×   ×   ×   +
7   8-1=7   4
=   =   =   =
2   3   2   1
1×2=2   1×3=3
```

012

```
3+0=3   6+2=8
×   ×   +   ×
6   5-1=4   4
=   =   =   =
1   1   1   3
8-3=5   0+2=2
  -       +
  3       4
  =       =
2+0=2   2+6=8
+   ×   ×   ×
9   9-1=8   7
=   =   =   =
1   1   1   5
1×8=8   6-0=6
```

013

```
4+0=4   1+7=8
×   ×   +   ×
3   7+2=9   2
=   =   =   =
1   2   1   1
2+6=8   0+6=6
  +       ×
  1       1
  =       =
2+7=9   9-6=3
×   ×   ×   +
5   5-1=4   7
=   =   =   =
1   4   3   1
0+5=5   6×0=0
```

014

```
7-1=6   2+4=6
×   +   ×   ×
3   5+4=9   3
=   =   =   =
2   1   1   1
1×1=1   8×1=8
  -       +
  1       1
  =       =
3-0=3   3+2=5
×   ×   ×   ×
4   6-1=5   2
=   =   =   =
1   1   1   1
2×4=8   5-5=0
```

015

```
9-2=7   4-1=3
×   +   ×   ×
6   3+5=8   7
=   =   =   =
5   1   3   2
4-4=0   2-1=1
  +       +
  1       5
  =       =
7-5=2   3+6=9
×   ×   ×   +
3   8+1=9   6
=   =   =   =
2   1   2   1
1+5=6   7-2=5
```

016

```
9-6=3   2-0=2
+   ×   ×   ×
3   6+2=8   8
=   =   =   =
1   1   1   1
2+6=8   6+0=6
  -       +
  1       1
  =       =
4+5=9   8+1=9
+   +   +   ×
7   9-5=4   3
=   =   =   =
1   1   1   2
1×8=8   2+5=7
```

017

```
3+0=3   8+1=9
×   ×   +   ×
4   6-3=3   9
=   =   =   =
1   1   1   8
2+6=8   1-0=1
  -       +
  6       1
  =       =
2+0=2   7+1=8
+   ×   +   ×
9   5+1=6   5
=   =   =   =
1   1   1   4
1-1=0   3-3=0
```

018

```
6+0=6   7-4=3
×   ×   ×   +
2   3×1=3   8
=   =   =   =
1   1   2   1
2×4=8   1+0=1
  +       +
  1       3
  =       =
4+5=9   6-3=3
×   ×   ×   ×
4   8×1=8   7
=   =   =   =
1   7   4   2
6-4=2   8-7=1
```

019

```
9-7=2   3-0=3
×   +   ×   +
3   8-2=6   7
=   =   =   =
2   1   1   1
7-7=0   8-8=0
  -       -
  7       8
  =       =
2+0=2   9+0=9
+   ×   ×   ×
9   5-1=4   8
=   =   =   =
1   1   3   7
1-1=0   6-4=2
```

020

```
1+3=4   9-0=9
+   ×   +   +
9   4-2=2   5
=   =   =   =
1   1   1   1
0+6=6   1×4=4
  +       +
  1       2
  =       =
2+7=9   9-6=3
×   ×   ×   ×
8   9-6=3   7
=   =   =   =
1   8   2   2
6-5=1   7-6=1
```

021

```
1+3=4   9+0=9
+   ×   +   ×
9   6-4=2   2
=   =   =   =
1   2   1   1
0+4=4   1×8=8
  +       -
  1       7
  =       =
2+5=7   5+1=6
×   ×   ×   ×
7   7-5=2   9
=   =   =   =
1   4   1   5
4+5=9   0+4=4
```

022

```
9-7=2   4+5=9
+   ×   +   ×
5   7×1=7   4
=   =   =   =
1   1   1   3
4-0=4   1×6=6
  +       -
  5       6
  =       =
8-5=3   8-0=8
×   ×   ×   +
4   8-3=5   8
=   =   =   =
3   1   4   1
2-1=1   0+6=6
```

023

```
2+3=5   7+0=7
×   ×   ×   ×
8   7-2=5   2
=   =   =   =
1   3   3   1
6-1=5   5-1=4
  +       -
  1       1
  =       =
8-2=6   1+0=1
×   +   +   +
5   5+4=9   9
=   =   =   =
4   1   1   1
0+1=1   0×5=0
```

024

```
9-5=4   3+1=4
+   ×   ×   +
5   3+4=7   7
=   =   =   =
1   1   2   1
4-2=2   1+0=1
  -       +
  2       4
  =       =
8-0=8   6-4=2
×   ×   ×   ×
5   5+1=6   7
=   =   =   =
4   4   3   1
0×9=0   6-2=4
```

025

9 − 2 = 7　7 − 0 = 7
+ × × +
7　2 + 1 = 3　8
1　1　2　1
6 − 2 = 4　1 + 4 = 5
× −
2　4
5 + 4 = 9　3 − 0 = 3
× + × +
2　3 + 1 = 4　9
1　1　1　1
0 + 2 = 2　2 × 1 = 2

026

4 + 0 = 4　2 + 7 = 9
× × + ×
8　7 + 1 = 8　2
3　2　1　1
2 × 4 = 8　0 + 8 = 8
+ −
1　7
4 + 5 = 9　4 − 1 = 3
× + × ×
8　3 + 4 = 7　7
3　1　2　2
2 + 0 = 2　8 − 7 = 1

027

7 + 0 = 7　4 − 0 = 4
× + × ×
3　5 + 4 = 9　3
2　1　3　1
1 + 1 = 2　6 − 4 = 2
− −
1　3
3 + 0 = 3　8 + 1 = 9
+ × + ×
8　7 − 4 = 3　4
1　2　1　3
1 − 0 = 1　1 × 6 = 6

028

9 − 6 = 3　9 − 6 = 3
+ × + ×
9　6 + 3 = 9　6
1　1　1　1
8 × 1 = 8　8 − 0 = 8
− +
1　2
5 + 0 = 5　9 − 2 = 7
+ × + ×
5　5 + 1 = 6　3
1　2　1　2
0 + 5 = 5　5 − 4 = 1

029

9 − 4 = 5　2 − 0 = 2
× + × ×
4　5 + 3 = 8　9
3　1　1　1
6 − 6 = 0　6 + 2 = 8
− −
6　1
5 − 0 = 5　6 − 1 = 5
+ + × ×
5　5 + 4 = 9　8
1　1　5　4
0 × 3 = 0　4 − 4 = 0

030

8 − 6 = 2　5 − 0 = 5
× + × ×
5　8 + 1 = 9　6
4　1　4　3
0 × 1 = 0　5 − 5 = 0
+ −
4　4
3 + 5 = 8　9 − 1 = 8
× + × +
9　4 + 1 = 5　2
2　3　4　1
7 − 5 = 2　5 − 5 = 0

031

6 − 2 = 4　5 + 3 = 8
× × × +
2　4 + 4 = 8　4
1　1　4　1
2 × 3 = 6　0 + 2 = 2
− −
2　2
8 − 1 = 7　4 + 0 = 4
× + + ×
5　3 + 4 = 7　9
4　1　1　3
0 × 4 = 0　1 × 6 = 6

032

7 − 0 = 7　7 + 0 = 7
× + + ×
3　9 − 1 = 8　3
2　1　1　2
1 + 5 = 6　5 − 4 = 1
− −
4　4
5 − 1 = 4　9 − 0 = 9
× + + ×
6　9 − 4 = 5　7
3　1　1　6
0 + 3 = 3　4 − 1 = 3

033

5 + 2 = 7　5 + 0 = 5
× + + ×
7　8 − 2 = 6　2
3　1　1　1
5 − 0 = 5　1 × 0 = 0
+ +
2　4
2 + 2 = 4　3 + 4 = 7
× × + ×
9　7 + 1 = 8　3
1　2　1　2
8 − 0 = 8　1 − 0 = 1

034

9 − 1 = 8　2 − 0 = 2
× + × +
8　4 + 1 = 5　9
7　1　1　1
2 + 0 = 2　0 + 1 = 1
+ +
2　4
9 − 2 = 7　2 + 5 = 7
× × × ×
9　7 + 2 = 9　4
8　4　1　1
1 × 9 = 9　8 − 7 = 1

035

5 + 4 = 9　5 + 0 = 5
× + + ×
7　5 + 4 = 9　2
3　1　1　1
5 − 1 = 4　4 − 4 = 0
+ −
4　4
4 + 5 = 9　9 − 0 = 9
+ × × +
7　3 + 4 = 7　9
1　2　6　1
1 + 6 = 7　3 + 5 = 8

036

8 + 0 = 8　4 − 0 = 4
+ × × +
2　4 + 4 = 8　9
1　3　3　1
0 + 2 = 2　2 + 1 = 3
+ +
3　4
7 − 5 = 2　3 + 5 = 8
× + + ×
3　9 − 2 = 7　5
2　1　1　4
1 × 1 = 1　0 × 2 = 0

037

```
8-3=5   3+5=8
×  +    +   ×
5  9-1=8    5
=  =    =   =
4  1    1   4
0+4=4   1×0=0
  −       +
  2       4
  =       =
9-2=7   5+4=9
×  ×    ×   ×
9  7-4=3    5
=  =    =   =
8  4    1   4
1×9=9   5-0=5
```

038

```
7-5=2   3+5=8
×  ×    ×   ×
7  7-3=4    5
=  =    =   =
4  1    1   4
9-5=4   2×0=0
  −       +
  4       4
  =       =
6-1=5   3+4=7
+  ×    +   ×
7  9-1=8    3
=  =    =   =
1  4    1   2
3+2=5   1-0=1
```

039

```
8-0=8   3-0=3
×  +    ×   +
4  4+3=7    9
=  =    =   =
3  1    2   1
2+0=2   1+1=2
  +       +
  3       3
  =       =
3+3=6   8-4=4
×  +    +   ×
8  8-4=4    3
=  =    =   =
2  1    1   1
4×1=4   2-0=2
```

040

```
4+2=6   7+0=7
×  +    +   ×
8  8-4=4    3
=  =    =   =
3  1    1   2
2×2=4   1-0=1
  −       +
  2       4
  =       =
2+0=2   3+4=7
+  ×    ×   +
9  8-4=4    7
=  =    =   =
1  1    1   1
1×6=6   2×2=4
```

041

```
9-6=3   2-0=2
×  +    ×   +
8  7-2=5    9
=  =    =   =
7  1    1   1
2-2=0   0+1=1
  +       −
  4       1
  =       =
3+6=9   5+0=5
×  +    ×   ×
9  8-4=4    8
=  =    =   =
2  1    2   4
7×1=7   0×4=0
```

042

```
8-4=4   9-0=9
×  +    ×   +
5  8+1=9    2
=  =    =   =
4  1    8   1
0+2=2   1-0=1
  +       +
  1       4
  =       =
3+3=6   8-4=4
+  ×    ×   +
9  7-4=3    6
=  =    =   =
1  4    2   1
2-0=2   4-4=0
```

043

```
9-1=8   4-0=4
+  ×    ×   +
6  2+3=5    9
=  =    =   =
1  1    2   1
5+1=6   0+3=3
  −       −
  1       3
  =       =
9+0=9   4+0=4
×  +    +   ×
4  3+3=6    5
=  =    =   =
3  1    1   2
6-4=2   0×9=0
```

044

```
9-3=6   2-0=2
×  +    ×   +
2  4+1=5    9
=  =    =   =
1  1    1   1
8-8=0   0+1=1
  −       −
  2       1
  =       =
9-6=3   8-0=8
+  +    ×   +
9  6×1=6    2
=  =    =   =
1  1    1   1
8×1=8   4×0=0
```

045

```
5+0=5   4+5=9
+  ×    ×   ×
9  8-4=4    3
=  =    =   =
1  4    1   2
4-4=0   6+1=7
  +       +
  3       3
  =       =
2+7=9   5+4=9
×  +    ×   ×
7  9×1=9    5
=  =    =   =
1  1    4   4
4×2=8   5×1=5
```

046

```
4+3=7   8-5=3
×  +    ×   +
4  9-4=5    9
=  =    =   =
1  1    4   1
6+0=6   0+2=2
  +       −
  2       2
  =       =
2+2=4   8-0=8
+  ×    +   ×
9  6-3=3    6
=  =    =   =
1  2    1   4
1×4=4   1+7=8
```

047

```
2+4=6   5+2=7
×  +    +   ×
9  9-3=6    3
=  =    =   =
1  1    1   2
8-3=5   1×1=1
  −       +
  2       3
  =       =
6-1=5   7-4=3
×  +    ×   +
5  8-2=6    9
=  =    =   =
3  4    4   1
0×7=0   2+0=2
```

048

```
6-2=4   5-1=4
×  +    ×   ×
2  6+3=9    8
=  =    =   =
1  1    1   3
2-2=0   4-2=2
  +       +
  4       4
  =       =
2+6=8   2+6=8
×  ×    ×   ×
5  5×1=5    5
=  =    =   =
1  4    1   4
0×6=0   0×9=0
```

049

```
5+0=5   8-3=5
+   ×   ×   +
9   2+3=5   8
=   =   =   =
1   1   4   1
4-4=0   0+3=3
  -       -
  4       3
  =       =
7+0=7   2+0=2
×   +   +   ×
3   9-1=8   5
=   =   =   =
2   1   1   1
1+5=6   0×4=0
```

050

```
8+0=8   7-5=2
+   ×   +   ×
2   5+3=8   5
=   =   =   =
1   4   1   1
0×1=0   5×0=0
  +       +
  4       2
  =       =
3+5=8   5-2=3
×   ×   ×   ×
9   5-1=4   7
=   =   =   =
2   4   2   2
7-7=0   0+1=1
```

051

```
8-0=8   7-2=5
×   +   +   ×
4   9-3=6   6
=   =   =   =
3   1   1   3
2+5=7   3-3=0
  -       +
  2       4
  =       =
9-3=6   9-7=2
×   ×   ×   ×
2   7+2=9   9
=   =   =   =
1   4   8   1
8-6=2   1+7=8
```

052

```
5+3=8   2+2=4
+   ×   ×   ×
9   5+1=6   9
=   =   =   =
1   4   1   3
4-4=0   2×3=6
  -       -
  4       3
  =       =
7+0=7   8+0=8
+   ×   +   ×
7   8-3=5   5
=   =   =   =
1   5   1   4
4+2=6   3-3=0
```

053

```
8+0=8   5+3=8
+   ×   ×   ×
3   7-4=3   5
=   =   =   =
1   5   1   4
1×6=6   5-5=0
  -       -
  4       4
  =       =
6+2=8   9-1=8
×   +   +   ×
5   5+3=8   5
=   =   =   =
3   1   1   4
0+3=3   7-7=0
```

054

```
8+1=9   8-0=8
+   ×   ×   +
2   6-1=5   6
=   =   =   =
1   5   4   1
0+4=4   0+4=4
  ×       -
  1       4
  =       =
9-4=5   4+0=4
×   +   +   ×
5   7+2=9   9
=   =   =   =
4   1   1   3
5-3=2   3+3=6
```

055

```
2+0=2   9-2=7
+   ×   +   ×
9   5-4=1   9
=   =   =   =
1   1   1   6
1-1=0   0+3=3
  ×       +
  1       4
  =       =
7-1=6   2+7=9
×   +   ×   ×
8   9-4=5   3
=   =   =   =
5   1   1   2
6-1=5   0+7=7
```

056

```
9-2=7   8-0=8
×   ×   ×   ×
7   7×1=7   8
=   =   =   =
6   4   5   6
3×3=9   6-2=4
  -       +
  3       1
  =       =
8-0=8   5+3=8
×   +   ×   ×
4   4+4=8   5
=   =   =   =
3   1   4   4
2+0=2   0×5=0
```

057

```
9-6=3   2-0=2
×   ×   ×   +
9   4+1=5   9
=   =   =   =
8   1   1   1
1×2=2   0+1=1
  +       -
  4       1
  =       =
7-6=1   7+0=7
×   +   ×   +
3   9-1=8   7
=   =   =   =
2   1   5   1
1-1=0   6-2=4
```

058

```
5+3=8   4+1=5
×   +   +   ×
7   7-1=6   6
=   =   =   =
3   1   1   3
5+0=5   0×4=0
  +       -
  1       4
  =       =
2+1=3   9+0=9
+   ×   ×   ×
9   6-4=2   9
=   =   =   =
1   1   1   8
1×8=8   8-7=1
```

059

```
8+0=8   4+5=9
+   ×   ×   ×
2   5+2=7   4
=   =   =   =
1   4   2   3
0×1=0   8-2=6
  +       ×
  4       1
  =       =
3+5=8   9-2=7
×   ×   +   ×
7   7+1=8   9
=   =   =   =
2   5   1   6
1×6=6   7-4=3
```

060

```
1+7=8   3+5=8
+   ×   +   ×
9   8×1=8   4
=   =   =   =
1   6   1   3
0+4=4   1+1=2
  -       +
  4       2
  =       =
7+0=7   9-3=6
×   ×   ×   ×
5   8×1=8   2
=   =   =   =
3   5   7   1
5+1=6   2+0=2
```

061

```
7+2=9   8+1=9
+   ×   +   ×
5   8-4=4   8
=   =   =   =
1   7   1   7
2-0=2   2-0=2
  ×       +
  1       4
  =       =
9+0=9   5+4=9
×   +   +   ×
2   9-2=7   8
=   =   =   =
1   1   1   7
8-0=8   2×1=2
```

062

```
8-1=7   3-0=3
×   +   ×   +
3   3+4=7   9
=   =   =   =
2   1   2   1
4×0=0   1+1=2
  +       -
  2       1
  =       =
5-2=3   7+0=7
+   ×   +   ×
8   6-2=4   3
=   =   =   =
1   1   1   2
3+5=8   1-0=1
```

063

```
7+0=7   2-0=2
+   ×   ×   +
9   6-1=5   9
=   =   =   =
1   4   1   1
6-4=2   0+1=1
  -       -
  4       1
  =       =
9-0=9   7+0=7
×   +   ×   +
4   5+3=8   7
=   =   =   =
3   1   5   1
6-2=4   6-2=4
```

064

```
8-0=8   5-0=5
+   ×   ×   +
3   5-3=2   9
=   =   =   =
1   4   1   1
1-1=0   0+4=4
  +       -
  3       3
  =       =
7-4=3   6-1=5
×   +   ×   +
3   9-2=7   9
=   =   =   =
2   1   4   1
1+1=2   2+2=4
```

065

```
4+2=6   8-7=1
×   ×   ×   +
8   3+2=5   9
=   =   =   =
3   1   4   1
2×4=8   0×1=0
  -       +
  4       3
  =       =
7+0=7   4+4=8
×   ×   +   ×
3   4+4=8   3
=   =   =   =
2   2   1   2
1+7=8   2+2=4
```

066

```
7-0=7   9-1=8
×   +   ×   ×
3   4+2=6   3
=   =   =   =
2   1   5   2
1+0=1   4-0=4
  +       +
  1       4
  =       =
6-1=5   4+4=8
×   ×   +   ×
5   6+3=9   4
=   =   =   =
3   3   1   3
0×5=0   3-1=2
```

067

```
8+0=8   6+2=8
+   +   +   ×
2   2+4=6   4
=   =   =   =
1   1   1   3
0×3=0   2-0=2
  -       +
  3       4
  =       =
2+0=2   4+4=8
×   ×   +   ×
9   9-1=8   4
=   =   =   =
1   1   1   3
8-0=8   2-0=2
```

068

```
9+0=9   3+5=8
+   ×   +   ×
7   4+4=8   6
=   =   =   =
1   3   1   4
6×1=6   1×8=8
  +       -
  4       4
  =       =
3+5=8   9-4=5
×   ×   +   ×
7   6-2=4   4
=   =   =   =
2   4   1   2
1×8=8   3-3=0
```

069

```
9-6=3   4+1=5
+   ×   +   ×
8   9-3=6   2
=   =   =   =
1   2   1   1
7×1=7   0×1=0
  +       +
  2       4
  =       =
8-3=5   3+5=8
×   ×   +   ×
5   8+1=9   3
=   =   =   =
4   4   1   2
0×6=0   2+2=4
```

070

```
9-3=6   2+5=7
×   +   +   ×
8   6+3=9   6
=   =   =   =
7   1   1   4
2+0=2   1+1=2
  +       +
  4       4
  =       =
6-4=2   4+5=9
+   ×   ×   ×
9   9×1=9   4
=   =   =   =
1   1   3   3
5+3=8   6×1=6
```

071

```
3+5=8   9+0=9
×   ×   +   ×
5   4+4=8   7
=   =   =   =
1   3   1   6
5-3=2   7-4=3
  +       -
  1       4
  =       =
6-4=2   8-0=8
×   ×   ×   +
2   8-2=6   2
=   =   =   =
1   1   4   1
2×3=6   8×0=0
```

072

```
9-0=9   9-0=9
×   +   ×   +
9   2+3=5   2
=   =   =   =
8   1   4   1
1+0=1   5-4=1
  +       -
  3       4
  =       =
6+3=9   3-0=3
×   ×   ×   +
8   6-2=4   9
=   =   =   =
4   5   1   1
8-4=4   2+0=2
```

073

1+7=8 7-0=7
\+ × × +
9 7-4=3 7
= = = =
1 5 2 1
0+6=6 1×4=4
\- -
4 4
= =
7+2=9 4-0=4
× × × ×
3 2+4=6 6
= = = =
2 1 2 2
1+7=8 4-0=4

074

7-1=6 8-0=8
× + × +
3 8-4=4 2
= = = =
2 1 3 1
1×4=4 2-2=0
× -
1 1
= =
3+4=7 8-1=7
× + + +
7 5+3=8 3
= = = =
2 1 1 1
1×2=2 6×0=0

075

8+0=8 4-0=4
\+ × × +
4 4+4=8 9
= = = =
1 3 3 1
2-0=2 2+1=3
\+ ×
1 1
= =
8+1=9 4+1=5
× + × ×
4 2+3=5 8
= = = =
3 1 2 4
2-1=1 0×4=0

076

8+1=9 8+0=8
\+ × + ×
4 9-1=8 4
= = = =
1 8 1 3
2-1=1 6-4=2
\+ -
4 4
= =
8-5=3 6-0=6
× + + ×
4 9-4=5 2
= = = =
3 1 1 1
2+0=2 1+1=2

077

9-5=4 5+3=8
× × × +
6 4+4=8 4
= = = =
5 1 4 1
4+2=6 0+2=2
\+ +
3 3
= =
2+5=7 8-5=3
\+ × × ×
9 3+3=6 9
= = = =
1 2 4 2
1-0=1 8-1=7

078

5+3=8 3+2=5
\+ × × +
9 5+4=9 5
= = = =
1 4 2 1
4-4=0 7×0=0
\- +
4 4
= =
9+0=9 5+4=9
\+ × + ×
8 8-1=7 8
= = = =
1 7 1 7
7-5=2 2×1=2

079

9-7=2 8-0=8
× + × +
5 8×1=8 2
= = = =
4 1 6 1
5-5=0 4-4=0
\+ -
2 4
= =
9-7=2 4-0=4
× × + ×
7 7+2=9 9
= = = =
6 1 1 3
3+1=4 3×2=6

080

7-1=6 4-0=4
× × × +
3 4+4=8 9
= = = =
2 2 3 1
1×4=4 2+1=3
\- +
4 2
= =
6+0=6 6+3=9
× + × ×
2 8-3=5 3
= = = =
1 1 3 2
2+2=4 0+7=7

081

2+6=8 9-7=2
\+ × × +
9 9-3=6 9
= = = =
1 7 5 1
1×2=2 4-3=1
\+ -
2 1
= =
6-4=2 2+2=4
\+ × × +
6 9×1=9 7
= = = =
1 1 1 1
2+6=8 8-7=1

082

8-1=7 5+4=9
\+ × + ×
8 8-1=7 8
= = = =
1 5 1 7
6-0=6 2-0=2
\+ +
3 3
= =
2+3=5 5+3=8
× × × ×
5 6-2=4 5
= = = =
1 3 2 4
0×5=0 0×6=0

083

8+0=8 5-0=5
× + × +
4 4+4=8 9
= = = =
3 1 4 1
2+0=2 0+4=4
\+ -
4 4
= =
3+4=7 5-0=5
× × + ×
7 7-1=6 7
= = = =
2 4 1 3
1+8=9 1+4=5

084

2+3=5 9-5=4
× × + ×
5 6-3=3 8
= = = =
1 3 1 3
0×1=0 2×1=2
\+ +
4 2
= =
8-5=3 8-3=5
× + × +
5 7-2=5 6
= = = =
4 1 4 1
0×3=0 0+1=1

085

```
8-2=6   3-0=3
×   +   ×   +
3   4+3=7   9
=   =   =   =
2   1   2   1
4×0=0   1+1=2
  +       -
  2       1
  =       =
9-2=7   3-0=3
×   ×   ×   ×
8   5+2=7   5
=   =   =   =
7   3   2   1
2+3=5   1×5=5
```

086

```
9+0=9   9-1=8
+   ×   ×   +
2   7+2=9   6
=   =   =   =
1   6   8   1
1+2=3   1+3=4
  +       -
  4       3
  =       =
9-6=3   5+0=5
×   +   ×   ×
9   9×1=9   9
=   =   =   =
8   1   4   4
1+1=2   5-0=5
```

087

```
5-0=5   8-3=5
×   +   ×   +
2   9-4=5   9
=   =   =   =
1   1   4   1
0+4=4   0+4=4
  -       -
  4       4
  =       =
6+0=6   2-0=2
×   ×   +   ×
7   5+4=9   8
=   =   =   =
4   3   1   1
2-2=0   1×6=6
```

088

```
7+2=9   8-5=3
+   ×   +   ×
6   9-3=6   8
=   =   =   =
1   8   1   2
3-2=1   4×1=4
  -       +
  1       4
  =       =
7-1=6   4+5=9
×   +   +   ×
3   8+1=9   7
=   =   =   =
2   1   1   6
1×4=4   3-0=3
```

089

```
3+3=6   3-0=3
×   +   ×   +
7   6-2=4   9
=   =   =   =
2   1   1   1
1×2=2   2+0=2
  -       +
  2       3
  =       =
9-0=9   5-3=2
×   +   ×   ×
8   3+4=7   8
=   =   =   =
7   1   3   1
2+0=2   5+1=6
```

090

```
5-0=5   9-4=5
×   +   +   ×
2   9-3=6   4
=   =   =   =
1   1   1   2
0+4=4   5-5=0
  -       +
  4       1
  =       =
7-0=7   3+6=9
×   +   ×   +
3   4+4=8   5
=   =   =   =
2   1   2   1
1+0=1   4-0=4
```

091

```
8-3=5   8-4=4
×   ×   +   ×
9   2+4=6   7
=   =   =   =
7   1   1   2
2×0=0   4×2=8
  +       -
  3       1
  =       =
2+3=5   8+1=9
×   ×   +   ×
8   4+4=8   9
=   =   =   =
1   2   1   8
6-6=0   6-5=1
```

092

```
7-4=3   7-1=6
×   ×   +   ×
7   9-4=5   6
=   =   =   =
4   2   1   3
9-2=7   2×3=6
  -       -
  2       2
  =       =
8-0=8   4-1=3
×   +   ×   +
2   3+2=5   9
=   =   =   =
1   1   2   1
6-5=1   0+2=2
```

093

```
6+3=9   4+4=8
×   +   ×   +
7   7+1=8   8
=   =   =   =
4   1   3   1
2+4=6   2×3=6
  -       -
  2       3
  =       =
5-2=3   5-0=5
×   ×   ×   +
7   8-3=5   6
=   =   =   =
3   2   2   1
5-1=4   5-4=1
```

094

```
8-0=8   9-0=9
×   +   +   +
9   8-4=4   7
=   =   =   =
7   1   1   1
2×3=6   3×2=6
  -       -
  2       1
  =       =
7-1=6   9-1=8
×   +   ×   +
4   7-2=5   2
=   =   =   =
2   1   4   1
8-5=3   5-5=0
```

095

```
7-1=6   3-1=2
×   +   ×   ×
3   9-1=8   7
=   =   =   =
2   1   2   1
1+4=5   4-0=4
  -       +
  3       2
  =       =
5-1=4   5-2=3
×   +   ×   +
6   6+3=9   8
=   =   =   =
3   1   4   1
0×1=0   5-4=1
```

096

```
2+0=2   7+2=9
+   ×   +   +
8   5-1=4   5
=   =   =   =
1   1   1   1
0×1=0   1×4=4
  +       ×
  2       1
  =       =
9-3=6   5+4=9
×   +   ×   +
9   7-1=6   7
=   =   =   =
8   1   3   1
1+2=3   0+6=6
```

097

```
7-6=1   8+1=9
×   +   ×   ×
2   9×1=9   2
=   =   =   =
1   1   7   1
4-4=0   2×4=8
  -       -
  2       3
  =       =
6-2=4   3-1=2
+   ×   +   ×
9   7+2=9   8
=   =   =   =
1   2   1   1
5+3=8   2+4=6
```

098

```
3+0=3   2+6=8
×   ×   ×   +
5   6×1=6   6
=   =   =   =
1   1   1   1
5+3=8   2×2=4
  -       -
  1       2
  =       =
7-2=5   3+0=3
×   +   +   ×
3   8+1=9   8
=   =   =   =
2   1   1   2
1+2=3   2+2=4
```

099

```
6+0=6   5+4=9
×   ×   +   ×
2   5+1=6   5
=   =   =   =
1   3   1   4
2-2=0   1+4=5
  -       -
  1       2
  =       =
5+1=6   2+2=4
×   +   ×   ×
7   9-3=6   7
=   =   =   =
3   1   1   2
5×1=5   2×4=8
```

100

```
3-0=3   6+3=9
+   ×   ×   ×
8   7×1=7   4
=   =   =   =
1   2   4   3
1+0=1   2×3=6
  +       -
  3       3
  =       =
9-3=6   8-0=8
+   ×   +   ×
4   7+1=8   5
=   =   =   =
1   4   1   4
3-1=2   6-6=0
```

101

```
6-4=2   6-1=5
+   ×   +   ×
9   8-4=4   4
=   =   =   =
1   1   1   2
5+1=6   0×3=0
  ×       -
  1       1
  =       =
2+1=3   5+2=7
×   ×   ×   ×
7   5-2=3   2
=   =   =   =
1   1   1   1
4+1=5   5-1=4
```

102

```
8-5=3   4-2=2
+   ×   ×   ×
3   4+4=8   9
=   =   =   =
1   1   3   1
1+1=2   2×4=8
  +       -
  3       4
  =       =
9-4=5   6+0=6
×   +   +   ×
7   8×1=8   8
=   =   =   =
6   1   1   4
3+0=3   4×2=8
```

103

```
3+4=7   9-6=3
×   ×   +   +
8   3-1=2   9
=   =   =   =
2   2   1   1
4-3=1   1×2=2
  -       +
  2       4
  =       =
6+1=7   8-6=2
+   ×   ×   ×
8   5-1=4   6
=   =   =   =
1   3   3   1
4+1=5   2-0=2
```

104

```
4+2=6   5+3=8
+   ×   +   ×
7   5+2=7   7
=   =   =   =
1   3   1   5
1-1=0   2×3=6
  +       -
  3       2
  =       =
4+4=8   5-1=4
+   ×   ×   ×
7   4+3=7   7
=   =   =   =
1   3   3   2
1+1=2   5+3=8
```

105

```
4+4=8   3+2=5
×   +   ×   ×
3   3+4=7   3
=   =   =   =
1   1   2   1
2-1=1   1×5=5
  +       -
  3       4
  =       =
9-4=5   2+1=3
+   ×   ×   +
5   3+3=6   9
=   =   =   =
1   1   1   1
4+1=5   2+0=2
```

106

```
8-5=3   8-0=8
+   +   +   ×
3   7-2=5   3
=   =   =   =
1   1   1   2
1×0=0   3+1=4
  +       +
  3       3
  =       =
2+3=5   4+4=8
×   ×   ×   ×
8   4+4=8   5
=   =   =   =
1   2   3   4
6-6=0   2-2=0
```

107

```
8-5=3   4+4=8
+   ×   ×   ×
5   4+4=8   2
=   =   =   =
1   1   3   1
3-1=2   2+4=6
  +       +
  3       1
  =       =
2+4=6   2+5=7
×   ×   ×   +
9   3+3=6   5
=   =   =   =
1   1   1   1
8-0=8   2×1=2
```

108

```
3+0=3   7-3=4
+   ×   ×   ×
8   8+1=9   8
=   =   =   =
1   2   6   3
1×4=4   3-1=2
  -       +
  4       4
  =       =
4-0=4   3+5=8
+   ×   ×   +
8   8-3=5   8
=   =   =   =
1   3   1   1
2-0=2   5+1=6
```

109

```
2+6=8   5+0=5
+   ×   ×   ×
9   3-1=2   4
=   =   =   =
1   2   1   2
1×4=4   0×3=0
  +       -
  2       3
  =       =
9-6=3   8+0=8
+   ×   +   ×
9   6-3=3   3
=   =   =   =
1   1   1   2
8×1=8   1+3=4
```

110

```
7+1=8   4-0=4
×   ×   ×   ×
6   6-3=3   6
=   =   =   =
4   4   1   2
2×4=8   2×2=4
  -       +
  3       1
  =       =
3+1=4   2+3=5
×   ×   ×   ×
6   8-1=7   6
=   =   =   =
1   3   1   3
8-6=2   4-4=0
```

111

```
7-2=5   1+7=8
+   ×   +   ×
3   9×1=9   3
=   =   =   =
1   4   1   2
0+5=5   0+4=4
  -       +
  3       1
  =       =
6-2=4   8-5=3
×   +   +   ×
4   6+2=8   5
=   =   =   =
2   1   1   1
4-4=0   6-1=5
```

112

```
8-5=3   4-0=4
×   ×   ×   ×
3   8+1=9   9
=   =   =   =
2   2   3   3
4×1=4   6×1=6
  -       ×
  1       1
  =       =
4+0=4   3+1=4
×   ×   ×   ×
4   9-2=7   4
=   =   =   =
1   3   2   1
6-0=6   1+5=6
```

113

```
8+0=8   8+1=9
+   ×   +   ×
6   5-2=3   4
=   =   =   =
1   4   1   3
4-4=0   1×6=6
  -       -
  4       3
  =       =
6+0=6   4+3=7
+   ×   ×   +
6   8-1=7   7
=   =   =   =
1   4   2   1
2×4=8   8-4=4
```

114

```
2+2=4   7-4=3
×   ×   +   +
6   7-3=4   7
=   =   =   =
1   2   1   1
2×4=8   1×0=0
  -       ×
  3       1
  =       =
7+1=8   3+0=3
+   ×   ×   +
7   3+4=7   9
=   =   =   =
1   2   2   1
4+0=4   1+1=2
```

115

```
8-0=8   3+6=9
×   +   +   ×
5   8×1=8   6
=   =   =   =
4   1   1   5
0+6=6   1+3=4
  -       +
  4       4
  =       =
6+2=8   2+7=9
+   ×   +   ×
5   5+3=8   3
=   =   =   =
1   4   1   2
1-1=0   0+7=7
```

116

```
6-2=4   5+4=9
×   ×   ×   +
5   5+4=9   3
=   =   =   =
3   2   4   1
0×3=0   5-3=2
  -       +
  3       3
  =       =
4+0=4   8-6=2
+   ×   ×   ×
9   3+1=4   6
=   =   =   =
1   1   3   1
3-1=2   2-0=2
```

117

```
5+4=9   9-3=6
×   +   ×   +
3   5+3=8   7
=   =   =   =
1   1   7   1
5-1=4   2+1=3
  +       -
  2       1
  =       =
7-3=4   2-0=2
×   ×   ×   +
6   7-2=5   8
=   =   =   =
4   2   1   1
2+6=8   0×6=0
```

118

```
7-3=4   3+1=4
×   +   +   ×
6   6+2=8   4
=   =   =   =
4   1   1   1
2-2=0   1×6=6
  +       -
  3       4
  =       =
8-5=3   6-2=4
×   +   ×   ×
5   7×1=7   6
=   =   =   =
4   1   4   2
0×1=0   2×2=4
```

119

```
7-0=7   4-0=4
×   ×   ×   ×
3   5+2=7   5
=   =   =   =
2   3   2   2
1+4=5   8×0=0
  -       ×
  1       1
  =       =
6+3=9   5-0=5
×   ×   ×   +
9   3+4=7   6
=   =   =   =
5   2   3   1
4+3=7   5-4=1
```

120

```
4+1=5   7+1=8
×   ×   ×   +
8   6+1=7   9
=   =   =   =
3   3   4   1
2-2=0   9-2=7
  +       -
  1       2
  =       =
8-3=5   4+0=4
×   ×   ×   +
4   9-1=8   6
=   =   =   =
3   4   3   1
2+3=5   2×0=0
```

121

```
8+0=8   9-5=4
+   ×   ×   +
2   5+2=7   9
=   =   =   =
1   4   6   1
0×1=0   3+0=3
  +       +
  4       1
  =       =
7-5=2   8+1=9
×   +   ×   ×
3   8-4=4   8
=   =   =   =
2   1   3   7
1-1=0   2+0=2
```

122

```
7+0=7   9-1=8
×   ×   +   +
3   2-1=1   2
=   =   =   =
2   1   1   1
1×4=4   0×2=0
  -       -
  4       1
  =       =
7+0=7   8-1=7
×   ×   +   ×
4   3+2=5   7
=   =   =   =
2   2   1   4
8-7=1   3×3=9
```

123

```
9-0=9   7-3=4
+   ×   +   ×
2   9-4=5   7
=   =   =   =
1   8   1   2
1-0=1   2×4=8
  +       -
  4       3
  =       =
3+4=7   9-1=8
×   +   ×   +
8   5+2=7   5
=   =   =   =
2   1   6   1
4-2=2   3+0=3
```

124

```
7+1=8   3+1=4
+   ×   +   ×
3   5+3=8   4
=   =   =   =
1   4   1   1
0×2=0   1×6=6
  -       -
  2       4
  =       =
2+0=2   7-2=5
+   ×   +   ×
9   7+2=9   4
=   =   =   =
1   1   1   2
1×4=4   6-6=0
```

125

```
8+0=8   4+3=7
+   ×   ×   ×
6   5-2=3   6
=   =   =   =
1   4   1   4
4-4=0   2-0=2
  -       +
  4       3
  =       =
3+0=3   5-3=2
×   ×   ×   ×
9   6-1=5   5
=   =   =   =
2   1   2   1
7+1=8   5×0=0
```

126

```
9+0=9   8-4=4
+   +   +   ×
2   3+1=4   6
=   =   =   =
1   1   1   2
1×2=2   2×2=4
  -       +
  1       4
  =       =
9-1=8   3+6=9
×   +   ×   ×
8   4+2=6   3
=   =   =   =
7   1   1   2
2+0=2   8-1=7
```

127

```
9-6=3   4-0=4
×   +   +   ×
9   9-2=7   4
=   =   =   =
8   1   1   1
1×2=2   1×6=6
  -       -
  2       4
  =       =
8+0=8   6-2=4
×   +   ×   ×
4   8×1=8   5
=   =   =   =
3   1   4   2
2+4=6   8-8=0
```

128

```
7+1=8   2+2=4
+   ×   ×   ×
5   5+1=6   7
=   =   =   =
1   4   1   2
2-2=0   2×4=8
  -       -
  2       4
  =       =
4+0=4   5-0=5
+   ×   ×   +
8   4×1=4   9
=   =   =   =
1   1   2   1
2×3=6   0+4=4
```

129

```
2+7=9   6+3=9
+   +   +   ×
9   4+2=6   8
=   =   =   =
1   1   1   7
1×3=3   2-0=2
  +       +
  2       2
  =       =
4+5=9   4+2=6
×   +   ×   ×
3   5+3=8   4
=   =   =   =
1   1   3   2
2+2=4   2×2=4
```

130

```
3+6=9   2+6=8
+   ×   ×   +
8   9-2=7   9
=   =   =   =
1   8   1   1
1×1=1   4+3=7
  +       +
  3       1
  =       =
2+4=6   9-4=5
+   ×   +   ×
9   3+3=6   7
=   =   =   =
1   1   1   3
1×8=8   5+0=5
```

131

```
2+6=8   5-0=5
+   ×   ×   +
9   4+4=8   9
=   =   =   =
1   3   4   1
1×2=2   0+4=4
  +       -
  1       4
  =       =
3+3=6   9+0=9
×   +   ×   +
8   6+2=8   4
=   =   =   =
2   1   7   1
4-2=2   2+1=3
```

132

```
2+1=3   5-0=5
×   ×   ×   +
9   6+2=8   9
=   =   =   =
1   1   4   1
8×1=8   0+4=4
  +       -
  4       3
  =       =
3+5=8   7+1=8
×   ×   ×   +
9   5+3=8   5
=   =   =   =
2   4   5   1
7-7=0   6-3=3
```

133

```
5+3=8   9-4=5
+   ×   ×   ×
9   5+4=9   4
=   =   =   =
1   4   8   2
4-4=0   1-1=0
  -       +
  3       4
  =       =
9-1=8   8-5=3
×   +   ×   +
2   4×1=4   9
=   =   =   =
1   1   3   1
8-6=2   2+0=2
```

134

```
7-1=6   5+3=8
×   ×   +   ×
4   9-2=7   2
=   =   =   =
2   5   1   1
8-4=4   2×3=6
  -       +
  3       2
  =       =
4+1=5   9-5=4
×   ×   +   ×
5   8×1=8   5
=   =   =   =
2   4   1   2
0×8=0   7-7=0
```

135

```
2+2=4   6+3=9
+   ×   +   ×
9   6+3=9   8
=   =   =   =
1   2   1   7
1×4=4   5-3=2
  -       -
  3       3
  =       =
4+1=5   8-0=8
×   ×   ×   +
5   8-4=4   4
=   =   =   =
2   4   3   1
0×1=0   2+0=2
```

136

```
7-0=7   7+0=7
+   ×   +   ×
8   8-4=4   3
=   =   =   =
1   5   1   2
5+1=6   1-0=1
  +       +
  3       3
  =       =
5+4=9   5-3=2
×   +   ×   ×
9   6-1=5   5
=   =   =   =
4   1   2   1
5×1=5   5×0=0
```

137

```
7+0=7   8-0=8
+   ×   ×   +
9   6+3=9   4
=   =   =   =
1   4   7   1
6-4=2   2-0=2
  -       +
  4       4
  =       =
4-0=4   5+4=9
×   +   +   ×
9   7+2=9   8
=   =   =   =
3   1   1   7
6-5=1   4-2=2
```

138

```
1+7=8   9-0=9
+   ×   ×   ×
9   2+4=6   7
=   =   =   =
1   1   5   6
0+6=6   4-1=3
  ×       +
  1       4
  =       =
8-6=2   9-5=4
×   ×   +   ×
5   5+4=9   4
=   =   =   =
4   1   1   1
0×8=0   8-2=6
```

139

```
3+6=9   9-0=9
+   ×   ×   +
9   6+3=9   3
=   =   =   =
1   5   8   1
2×2=4   1+1=2
  +       -
  2       1
  =       =
2+4=6   4-0=4
+   ×   +   ×
9   3+4=7   4
=   =   =   =
1   1   1   1
1×8=8   1×6=6
```

140

```
6+1=7   8-3=5
×   +   ×   +
9   8-3=5   6
=   =   =   =
5   1   4   1
4+1=5   0+1=1
  +       ×
  4       1
  =       =
3+5=8   9-1=8
×   +   ×   +
7   7-2=5   2
=   =   =   =
2   1   4   1
1×5=5   5-5=0
```

141

```
5-3=2   8-7=1
×   +   ×   +
9   8-4=4   9
=   =   =   =
4   1   3   1
5×0=0   2-2=0
  +       -
  1       2
  =       =
9-1=8   7-0=7
×   +   ×   ×
9   8-2=6   7
=   =   =   =
8   1   4   4
1×6=6   2+7=9
```

142

```
6+3=9   9-3=6
+   ×   +   ×
9   9-2=7   2
=   =   =   =
1   8   1   1
5-4=1   6-4=2
  -       -
  4       1
  =       =
8-0=8   9-3=6
+   ×   +   ×
9   5+4=9   7
=   =   =   =
1   4   1   4
7-7=0   8-6=2
```

143

```
8-3=5   7+0=7
×   +   +   ×
5   6+2=8   3
=   =   =   =
4   1   1   2
0+1=1   5-4=1
  +       -
  4       4
  =       =
3+5=8   9-0=9
+   ×   ×   +
9   4+3=7   3
=   =   =   =
1   3   6   1
2-0=2   3-1=2
```

144

```
9-5=4   5+0=5
×   ×   +   ×
9   4+4=8   2
=   =   =   =
8   1   1   1
1×6=6   3-3=0
  -       -
  4       1
  =       =
8-2=6   2+2=4
×   ×   ×   ×
3   4+4=8   9
=   =   =   =
2   2   1   3
4-0=4   6×1=6
```

145

```
7+0=7   7+0=7
+   ×   ×   +
9   3+2=5   3
=   =   =   =
1   2   3   1
6-5=1   5-5=0
  -       +  
  4       1  
  =       =  
8+1=9   9-6=3
×   +   +   ×
5   8×1=8   9
=   =   =   =
4   1   1   2
0+7=7   7-0=7
```

146

```
5-0=5   5+4=9
×   +   +   ×
2   6+3=9   8
=   =   =   =
1   1   1   7
0+1=1   4-2=2
  +       -  
  3       1  
  =       =  
8-4=4   7+1=8
×   ×   +   ×
4   7-4=3   4
=   =   =   =
3   2   1   3
2×4=8   0+2=2
```

147

```
3+4=7   9-7=2
×   +   ×   +
4   9-1=8   9
=   =   =   =
1   1   7   1
2+4=6   2-1=1
  -       +  
  4       3  
  =       =  
8+0=8   8-4=4
+   ×   ×   +
4   9-4=5   9
=   =   =   =
1   7   4   1
2×1=2   0+3=3
```

148

```
9-1=8   6-2=4
×   +   ×   ×
9   4+3=7   7
=   =   =   =
8   1   4   2
1+1=2   2×4=8
  +       -  
  3       3  
  =       =  
2+4=6   5-1=4
×   +   ×   ×
5   4-2=2   5
=   =   =   =
1   1   1   2
0×7=0   0×9=0
```

149

```
2+7=9   6-4=2
+   +   +   ×
9   4+1=5   8
=   =   =   =
1   1   1   1
1×3=3   1×6=6
  +       -  
  1       3  
  =       =  
4+4=8   7-3=4
+   ×   +   ×
9   5+4=9   8
=   =   =   =
1   4   1   3
3-3=0   6-4=2
```

150

```
2-0=2   8-1=7
×   +   ×   +
8   9-4=5   5
=   =   =   =
1   1   4   1
6-5=1   0+2=2
  -       -  
  4       2  
  =       =  
7+1=8   3-0=3
×   +   ×   +
5   3+1=4   9
=   =   =   =
3   1   1   1
5-4=1   2+0=2
```